Notes

For those who may find this publication of benefit while preparing their research proposals, they may also consider getting in contact with Scientific Management Network S.L., an advisory company helping many researchers in the process of preparing their research proposals.
You can get in contact by phone at +34 698 707 986 or email at info@s-n-m.eu.

* In the interest of aiding those currently working on research proposals, Dr. Pascal Kahlem is offering a complimentary 30-minute consultation to anyone who purchases the four volumes of this book.
To avail of your free consultation, simply call or email Dr. Pascal Kahlem and simply mention the purchase code for each volume that you purchased.

Published by Pascal Kahlem

First edition: 2019

ISBN: 978-84-09-08158-5

Cover photo credit: Pascal Kahlem

About the author: Pascal Kahlem (PhD)

Owner of Scientific Network Management SL (Spain registered).

PhD in Human Genetics at the University Paris VII (France) obtained with honours in 1999. I pursued three Post-docs on Human Genetics successively at the CNRS (Paris, France), Max-Planck-Institute (Berlin, Germany) and Charité Hospital (Berlin, Germany). My research contributed to a better understanding of 1) the molecular mechanisms underlying neurodegeneration in Huntington's disease; 2) the molecular consequences of trisomy 21; 3) the possibility to induce cell arrest to stop lymphoma's progression.

Since 2006 I have been involved in operations management of large European Life Sciences research networks: ENFIN, SYBARIS, ProteomeXchange, MICROME, TransPLANT, and lately ELIXIR a Pan-European Infrastructure for Life Sciences, where I coordinated successively the Computing and Tools technical platforms and the Training platform. These international research networks aim at improving research quality and standards by bringing top European researchers together to focus on the most challenging research topics of our time such as personalised medicine, cancer, etc.

Scientific Network Management S.L. was created in 2013 to provide support to researchers and entrepreneurs on grant writing, assessment, training, and management of research and innovation projects.

WEB: http://www.grantwriter.eu

CONTACT: info@s-n-m.eu

LinkedIn Profile: https://www.linkedin.com/in/pascalkahlem

Selected publications:

- Transcript level alterations reflect gene dosage effects across multiple tissues in a mouse model of Down syndrome. Genome Res. 2004 Jul;14(7):1258-67.
- A gene expression map of human chromosome 21 orthologues in the mouse. Nature. 2002 Dec 5;420(6915):586-90.
- Peptides containing glutamine repeats as substrates for transglutaminase-catalyzed cross-linking: relevance to diseases of the nervous system. Proc Natl Acad Sci U S A. 1996 Dec 10;93(25):14580-5.

Table of contents

This book is meant to be a quick guide that every researcher and grant writer can keep at hand, especially at the time to start writing a grant proposal. It is based on the experience acquired by providing consulting support to more than three hundred researchers across more than 25 countries worldwide.

It has been kept concise and straight to the point, intending to focus on the main points that can help researchers to design their project and successfully pitch it to funding bodies.

This guide helps defining, understanding and structuring a research project. Even though this book is meant at guiding the writing of successful grant proposals for research projects, the tips provided will also be very useful to researchers to design their research projects, even if they do not aim to submit a grant proposal.

It is assumed that the success of a grant proposal does not only depend on the structure of the project, but also on its innovative contents and objectives. For this reason, we dedicate a complete volume on strategy building.

The book is divided in four volumes, each of them addressing a different aspect of the process of project design and grant writing.

Volume 1. The basics of grant writing
Defining a project
Life cycle of research projects

Volume 2. Project design
Project strategy building
Project reverse designing

Volume 3. Project writing
Excellence
Impact
Implementation
Budget planning

Volume 4. Additional considerations on grant writing
Evaluation criteria
Final tips

Volume 1

A project is an individual or collaborative enterprise that is carefully planned to achieve a particular aim. (Oxford Dictionaries)

Project's characteristics

A project intended to be presented to funding bodies, would generally include four main characteristics, each of which are detailed below:

1. Novelty / Innovation:

Novelty, or Innovation, usually refers to the translation of an idea into a commercially viable product. Although this book is focused on research and scientific projects, it may be helpful to look at the uniqueness of projects in a business sense, as the aim here is to help researchers in successfully applying for funding for their projects. In order to be successful in attaining grants or funding for a research project, it is useful to envision real world applications or impact, which may bring the project into the realm of commercial activity.

When we talk about novelty or innovation in research projects, it usually derives from a sound intellectual process and careful analysis of an unexpected outcome in within a project.
It is important to note that in an industrial context, promoting novelty or innovation in research is about having a working environment that is favourable for creative thinking. (de Souza, 2010, p.72,76)

2. Concretization / Formalization:

Concretizing consists of converting the ideas and the concept into a series of specific actions that will lead to the achievement of the project.
Formalizing consists in defining the roles of every item within the project's system. (O'Leary, 2017, Ch.1)

3. Will / Motivation:

Although there may be many clogs in the clockwork of a research product, none are quite as fundamental as the willpower to see the project through.

Effective motivation requires not only arousal or energy but also guidance by an effective and cognitive system that, at least for most of us, is susceptible to distraction or depletion. (Ryan, 2013, Ch.1)

4. Finality/Goal:

The finality or goal of a project is the expected outcome or result from the project. For a project goal to be successful it must have certain characteristics such as being:

- Specific: A wise project manager avoids ambiguity. What is intuitively obvious to one stakeholder is foreign to another.

- Measurable and verifiable: A wise project manager not only achieves a goal, but also validates it.

- Complete: All goals should be defined before starting a project.

- Consensual: All stakeholders should agree to the goals before it is expended. Agreements and expectations should be set before committing significant amounts of time and money to a project.

- Realistic: The project manager should make sure that all goals are achievable before starting the project.
(Bender, 2004, P.17,18)

Projects types

Projects can come in many different forms and areas of work, some of these may include organizing a conference, the development of a new invention, the setting up of a new financial system, the reorganization of a service and, of course, developing a research project, among others.

1. What is a research project?

When undertaking your research project, it is important to understand that research can have several legitimate objectives, either singly or in combination. The main, overriding objective must be that of gaining useful or interesting knowledge. (Walliman, 2011, P.7)
Research is the process of gathering data in order to answer a particular question and this question will generally relate to a need for knowledge that can facilitate problem solving. Research can help us:

- Understand more about particular issues and problems, including all the complexities, intricacies and implications thereof;
- Find workable solutions - vision futures, explore possibilities;
- Work towards that solution - implement real change;
- Evaluate success - find out if problem solving/change strategies have been successful;
- Offer robust recommendations - as an extension of findings, recommendations can be used to influence practice, programmes and policy.

(O'Leary, 2017, sect,1)

Another important factor that will determine the research project is the nature of its objective. Research can have many varying objectives such as categorization, explanation, prediction, creating a sense of understanding, providing potential for control and evaluation. (Walliman, 2011, P.7)

2. Types of research projects

As mentioned above, there are many types of projects when we talk about projects in general, however there are also many different types of projects, specifically within the field of research projects.
These different types include laboratory, literature, meta-analysis, intervention, questionnaire and data handling, among others.

Laboratory projects are typically based in a laboratory environment. The types of projects that are typically done will have some elements of repetition, sample preparation and analysis, for example measuring glucose in provided urine samples in order to accept or reject a hypothesis.

Literature projects review existing studies by collating data and conclusions to create a consensus dataset and conclusion. Sometimes this type of project may be viewed as less worthy than other types of research projects, especially when there is little or no data manipulation or analysis.

Meta-analysis projects are literature projects with complex models applied to reach a conclusion. These projects, by having data manipulation and analysis have considerable research currency.

Intervention projects are when the researcher recruits volunteers to take part in a piece of research. For example, taking vitamin C tablets for six weeks and then providing urine samples to analyse possible changes.

Questionnaire projects involve collection of data from volunteers rather than samples and are lower risk than intervention projects, but still require ethics and recruitment. A typical project could be a food frequency questionnaire to determine nutrient intake in a cohort.

Data handling projects are lower risk as the data will already have been obtained from a previous study and using statistical tests, hypotheses are tested. An example project may be to look at a case control data from a prostatic cancer study of 10,000 men, which contains data on cancer marker concentration, symptoms and lifestyle. (Basten, 2011, P.12,13,14)

II. Life cycle of research/scientific projects

To design a research project, you will need to start by defining the question that demands an answer or the need that requires a resolution or a riddle that seeks a solution, which can be developed into a research problem; the heart of the research project. (Walliman, 2011,P.29)

Once you have identified your unique research problem, the next step is to begin forming the project proposal that will later be used to apply to a funding body.

Research proposals often follow a defined pattern with common features within the proposal; an explanation, compact and precise fashion, of the nature of the research, why it is needed, how it will be done, the likely outcomes, their impact and, in most cases, exactly what resources are required to carry it out.

Assuming that the research proposal is successfully funded, the next step is to carry out the research and execute on the project. During the execution and at the end of the project, it is key for the project managers to report progress and results to the funding body and to the project director, in order to ensure that the progress is in line with the expectations.

Finally, the research project can take at least two paths: results dissemination and/or results exploitation. The intent of disseminating research is to spread

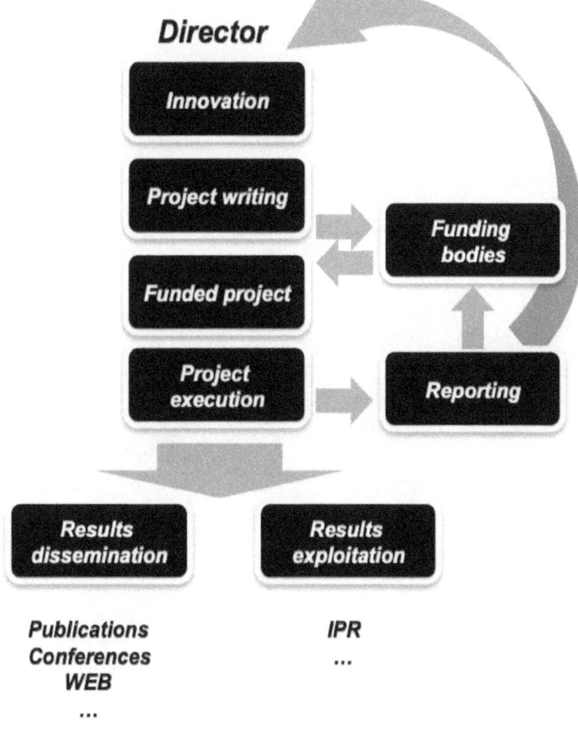

knowledge through publications, conferences, the internet, building understanding among various stakeholders, such as other scientists, industry, the public, policy makers, among others.
(Brownson, Colditz, Proctor, 2012,)
With regards to results exploitation, commercialization of project products will require the establishment of the Intellectual Property Rights between the project partners, and the development of an appropriate business plan.
(Siota, 2017).

III. Final Tips

- Envision real world applications, potentially bringing the project into commercial activity

- Start by defining the question that demands an answer

- Identify all the players necessary to perform the project until the goal is achieved

- Adjust the scope to the budget available

1. Oxford Dictionaries
2. Innovation in Industrial Research by Paulo Antonio de Souza (2010)
3. Merriam Webster
4. Introduction to Scientific Research Projects By Graham Basten (2011)
5. Your Research Project: Designing and Planning Your Work By Nicholas Walliman (2011)
6. Dissemination and Implementation Research in Health: Translating Science to Practice: by Ross C. Brownson, Graham A. Colditz, Enola K. Proctor (2012)
7. European Commission
8. Linked Innovation: Commercializing Discoveries at Research Centers By Josemaria Siota (2017) P.74,75
9. The How to Project Manage Series: Setting Goals and Expectations By Michael B Bender, Pmp (2004)
10. The Essential Guide to Doing Your Research Project By Zina O'Leary - Chapter 1.
11. The Oxford Handbook of Human Motivation edited by Richard M. Ryan (2013)

Grant writing = solution selling

As scientific research projects turn more and more into product development projects, funding bodies consider projects as business opportunities, where research aims at a product that can be exploited commercially.

Scientists are often not prepared to consider exploitation strategies in their projects. However it is with this mindset that they must tackle the preparation of grant proposals.

The following questions are meant to address the points that should be taken into account at the time to prepare a scientific project. The 6 points below are not exhaustive since all projects are different, but we hope that it will help you to clarify your approach leading to the writing a successful proposal. The following questions are extracted from the Project Strategy Builder interface that we offer online:
http://www.grantwriter.eu/psb

1. Project definition and development

- Define the purpose of the project and the reason to develop that product.
    General objective and Specific objectives necessary to reach the general objective.

- Define a vision (5-10 years) for the project and for the product that you want to achieve.

- What are the triggers that motivate the development?
    Prototype has reached maturity?
    A new technology is available that enables this development?
    A new demand has arisen?

- Identify the assets of the project and its impact.
    What are the needs of the users?
    Is your product matching the needs of the users?
    What is (are) the unique feature(s) that the competitors do not have?
    What impact does the product have at the socio and economical level, locally and internationally?

2. Market study

- Know your competitors.
    Is there already a similar product on the market?
    Time to create or revive partnerships?

- What is the product target group?
    This will define the design of the product. By defining your target customer, you can focus exactly on the needs that mean the most to those customers and divert attention away from product characteristics and service activities that are either not wanted or unappreciated. It is useful to consider that generally, companies select segmentation variables, including gender, education level, and income to define a target market. Companies determine the nature and number of variables to be used for defining target markets. Market segmentation benefits companies by;
    - *Permitting identification of customers,*
    - *Designing products and services to meet customer's needs,*
    - *Allowing them to develop effective promotions to stimulate demand.*
        (Duchessi, 2004, P.64)

- Estimate the Return on Investment.

3. Funding sources

- Assess the advantages / disadvantages of National, EU grant or international funding bodies.

- Choose the most appropriate funding calls.
    Submission deadline, constraints on consortium composition, topic, etc

4. Is lobbying an option?

- Program committees
- Advisory groups
- National/European/International technology platforms
- Public consultations
- Letters of support / interest from future users

5. Building a strong a multidisciplinary consortium

- Should you be the coordinator?
    Evaluate the return on investment of your own time.

- Gather trustable collaborators.
    Balance scientific, technical, political contributions and gender

6. Tasks distribution and budget planning

• Distribute the tasks amongst the partners.

• Propose a budget to each partner according to its contribution and needs, also maintaining a balance between all the partners of the consortium.

II. Project components

Once the project strategy is set, you need to build the different parts of the grant proposal.

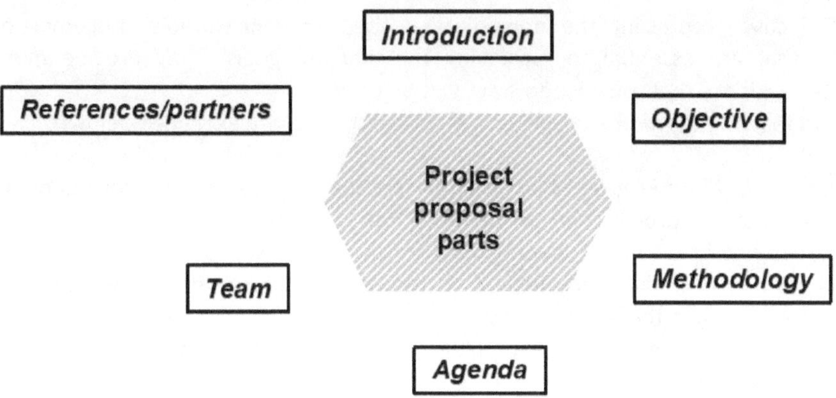

As we can see from the above diagram, a project proposal is comprised of several parts: the introduction, the objective, the methodology, the agenda, the team and the partners.

1. Introduction / Cover letter

The main purpose of the introduction or cover letter is to pitch the proposal to the reader.
A cover letter may include the following: the context, the goal, the key features of your proposal, the expected impact, the schedule quoted in the proposal and other key points.
The point of a proposal is to convince the funder that the proposed project is worth the investment.
(Coombs, 2005, P.18,19)

Early in the proposal, often in the introductory paragraph, it is wise to set forth an explicit statement of your purpose in undertaking the study. We use the word purpose in its general sense as a statement of why you want to do the study and what you intend to accomplish. Such statement can be divided broadly into those related to the desire to improve something and those reflecting a desire to understand something. In addition to such practical and theoretical purposes, Maxwell (2013) has pointed out that, in some instances, it may be wise to be explicit about more personal purposes as well, including interests related to simple curiosity, a sense of social responsibility or career demands.

A statement of purpose needs not be an exhaustive survey of your intentions, nor need it be written in the formal language of research questions (which are much more specific expressions of what you want to learn). (Locke, Spirduso, Silverman, 2014, P.9)

2. Objectives

Objectives represent the immediate desired and measurable outcomes or results that are essential for achieving the ultimate goals. They provide more tangible evidence that the desired state was achieved.

The two major types of objectives, process and outcome, are explained below:

- *Process objectives*: Process objectives (a) describe the expected improvements in the operations or procedures, (b) quantify the expected change in the usage of services or methods, or (c) identify how much service will be received. Process objectives do not indicate the impact on the program recipients. Rather they are formulated because the activities involved in implementation are important to the overall understanding of how a problem or need gets addressed. They help to provide insight into experimental, unique and innovative approaches or techniques used in a program.

- *Outcome objectives*: The second, and more common type, of objective is known as an outcome objective. An outcome objective specifies a target group and identifies what will happen to them as a result of the intervention or approach.
 (Coley, Scheinberg, 2008, P.48,50)

3. Methodology

There should be a transparent and obvious correlation between the approach and design / instruments / techniques so that they make sense as a system within a particular paradigm. The description and justification of that system is generally termed the methodology and this is followed by or includes a description of the planned research implementation (laboratory, fieldwork, documentary analysis or creative art, and so forth).
(Pam Denicolo, Lucinda Becker, 2012, P.62)

Although the basic logic of scientific methodology is the same in all fields, its specific techniques and approaches will vary, depending on the subject matter.
(Kumar, 2014, P.Ch.2)

4. Agenda / Timeline

A research timeline (or timetable) should chronologically list the major phases of the research project, along with the key tasks to be completed in each phases. Typically, start and finish dates for the overall project are indicated, as well as perhaps start and finish dates for each main phase. The timetable can either be set out as a list of main headings and sub-headings or as a table. If a table is used, summarize the start and finish dates for major sections of the project in the accompanying text.
When devising a timeline, one approach is to work backwards from the planned completion date for the project (where known). This involves estimating the amount of time likely to be required to complete the last task on the list, then counting backwards (e.g. in days or weeks) to the required starting time for that task. This process is then repeated for each task further up the list, until the start and finish times for all tasks are calculated.
(Thomas, Hodges, 2010, P.57,58)

5. Team

This section of the proposal focuses on introducing the team involved in pursuing the project, such as background information and qualifications.

A part of most grant proposals is a curriculum vitae, a biosketch, a bio-paragraph or a resume that identifies the key personnel, including the Principal Investigator (PI); and their education, professional experience, and publication history.

Background information consists of your and other researchers relevant previous research that directly relates to your proposed research. You explain this previous research in order to place your proposed research into a historical research context and to provide a backdrop against which to claim novelty and significance for your proposed research.
(Oster,Cordo, 2015, P.8,9)

6. References / Partners

This section refers to any external partners associated to the project such as funding partners or research partners.
In this section, provide a list of all external partners involved in the project and the extent of their involvement (what their exact role is, what they bring to the project, and what they expect to receive form the project).

Once the project strategy is established, it has to be turned into a project delivery plan, which will be delivered as one of the initial steps into the project. Project delivery planning is a method to maximize success by planning to manage all areas of uncertainty. Every project requires a robust project delivery plan if it is to be successfully carried out and the benefits realized sustainably:

> The project delivery plan is the link between what we have committed to delivering and what we actually do deliver.

The formal project delivery plan would typically be delivered prior to any actual project delivery, as it is considered to be:

- A formal agreement between the project sponsor and the project manager - effectively linking the project to the business.

- A communication and management tool for the project manager and project team effectively setting up the internal workings of the project.

- A communication tool between the project manager / project team and the external project stakeholders - further integrating the project with the organization in which it sits.
(Melton, 2011, P.11,12)

Reverse designing

Project reverse designing is the process of deciding on the goal(s) to be reached and time for accomplishment and, from there, designing the timeline of the project.

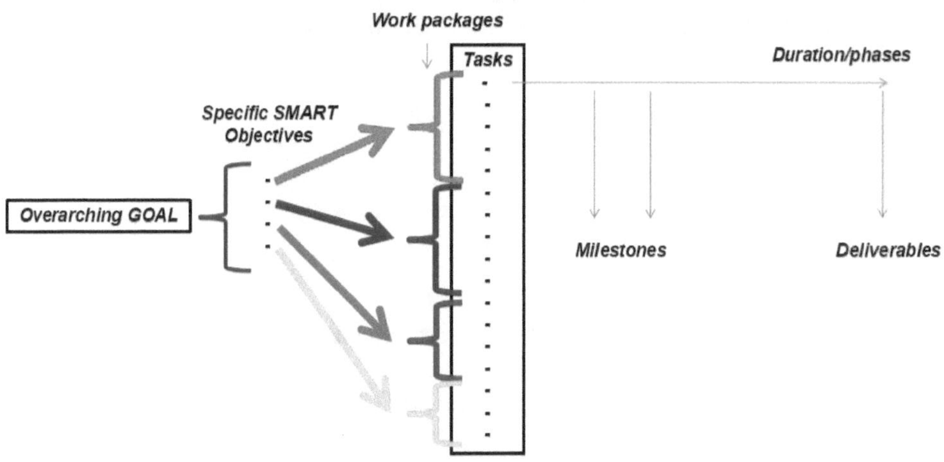

1. Overarching goal

Goals respond to identified needs or problems and are statements of the ultimate mission of the program. They represent an ideal or hope for state of the desired change. Most proposals identity one to three goals.
Goals are usually written indicating the system in which the services are to be provided. To write the goals, return to the needs or problems you seek to address and state the major changes that your work will produce.
(Coley, Scheinberg, 2008,P.47,48,49)

2. SMART objectives

The "smart" acronym refers to five concepts that must constantly be referred to when setting objectives, in order to validate their relevance. In order, the concepts are specific (S), measurable (M), assignable (A), realistic (R) and time bound (T).
This technique can also be used individually (by setting smart personal objectives) or in a team (a manager can set objectives for the group to achieve together).
Since we can define a goal as the result of a series of objectives that must be reached, they themselves can be divided into a series of sub-objectives. Some advantages of the SMART model are included below.
- Firstly, the model promotes achievement of concrete results by focusing on the tangible and quantifiable aspects of objectives.
- Secondly, it can be applied to various fields.
- Finally, the SMART criteria make the objective complete and require little or no additional details.
("50MINUTES", 2015, P.6,7,9)

3. Work packages

A work package is simply a low level task or job assignment. It describes the work to be accomplished by a specific performing working group or organization.
Work package is the generic term used in the criteria to identify discrete tasks that have definable end results.
Depending on the funding body, it is sometimes not necessary that work package documentation contain complete, stand-alone descriptions.

Short-term work packages may help evaluate accomplishments. Work packages should be natural subdivisions of effort planned according to the way the work will be done. However, when work packages are relatively short, little or no assessment of work-in-process is required and the evaluation of status is possible mainly on the basis of work package completions. The longer the work packages, the more difficult and subjective the work-in-process assessment becomes unless the packages are sub-divided by objective indicators such as discrete milestones with pre-assigned budget values or completion percentages.

(Kerzner, 2009, P.437)

4. Milestones, phases & deliverables

For processing a project successfully, it is necessary to sub-divide its execution into several phases, e.g. according to key events representing milestones (cf., e.g., Shutb et al. (1994, pp. 304)). The end of each phase or the time point of reaching a milestone provides a checkpoint at which the progress of the project can be controlled by its stakeholders, the assumptions of the plan may be verified, a modified plan may be developed, and its specifications may be refined. For this purpose, due dates when to finish a phase or to achieve a milestone are defined.
Having developed the milestones of a project, these are usually arranged in a milestone plan or Gantt chart, which is a directed graph showing the logical dependencies between the milestones, i.e., milestone b cannot be achieved before milestone a.
(Klein, 1999, P.18)

Management of project objectives is first established against deliverables related to the project life cycle. To do that, it is necessary to divide the total project time period - the life cycle - into a number of identifiable sections. The result is that what could be a relatively long time block is now presented as a sequence of time chunks or phases. These phases can be further sub-divided into stages.
Each phase is marked by the production of one or more deliverables, which might include; reports, design drawings, documentation, appointing a supplier or vendor, testing a package of software or commissioning an item of equipment.

The deliverables will vary with the industry or business. They result from a process, which could itself have contained milestones. For instance, if we were considering the surgical process, the pre-operation equipment checks, patient preparation, operation completion and patient consciousness could be the deliverables of the different phases. Deliverables from a preceding phase are usually approved before work can start on the next phase. When a later phase is commenced before the completion of a preceding phase (because the risks involved have been deemed acceptable), that is called fast-tracking. Each project phase normally includes a set of predefined work products designed to establish the desired level of management control.
Phases and deliverables are part of a generally sequential logic designed to ensure proper definition of the product of the project.
(Hamilton, 2001, P.84,85)

- Put be a heavy focus on the innovative contents and objectives.

- Reverse designing is deciding on the goal(s) to be reached and time for accomplishment and, from there, designing the timeline of the project

- Remember that the project delivery plan is the link between what was committed and what is delivered.

- To write the goals, return to the needs and state major changes that will be produced by the work

- Short-term work packages may help evaluate accomplishments.

- For processing a project successfully, it is necessary to sub-divide its execution into several phases.

- Divide the project the life cycle into identifiable sections to present as a sequence of time chunks or phases.

V. References

1. Real Project Planning: Developing a Project Delivery Strategy by Trish Melton (2011) P.11,12

2. IT Project Proposals: Writing to Win by Paul Coombs (2005) P.18,19

3. Proposals That Work By Lawrence F. Locke, Waneen Wyrick Spirduso, Stephen J. Silverman (2014) P.9

4. Proposal Writing: Effective Grantsmanship by Soraya M. Coley, Cynthia A. Scheinberg (2008) P.47,48, 49,50

5. Developing Research Proposals by Pam Denicolo, Lucinda Becker (2012), P.62

6. Research Methodology: A Step-by-Step Guide for Beginners by Ranjit Kumar (2014) Ch.2

7. Successful Grant Proposals in Science, Technology, and Medicine: A Guide to Writing the Narrative by Sandra Oster, Paul Cordo (2015) P.8,9

8. Designing and Managing Your Research Project: Core Skills for Social and Health Research by David Thomas, Ian D Hodges (2010) P.57,58

9. SMART Criteria: Become more successful by setting better goals by 50MINUTES.COM, (2015) P.6,7,9

10. Project Management: A Systems Approach to Planning, Scheduling, and Controlling by Harold Kerzner (2009) P.437

11. Scheduling of Resource-Constrained Projects by Robert Klein (1999) P.18

12. Managing Projects for Success: A Trilogy by Albert Hamilton (2001) P.84,85

13. Crafting Customer Value: The Art and Science by Peter Duchessi (2004) P.64]

Volume 3

Smart Objectives

Before writing your objectives, there are a number of questions that should be considered to help clarify what the desired achievement is from the project;

-What do you actually want to achieve,
-Why does it matter,
-Is this a new area of research or has it already been done? If it has been done
-before, what will be the benefit of you doing it again?
-Exactly what do you plan to do? And why?,
-If it doesn't work, what will you do next?,
-If it does work, what will you do next?,
-How does this fit in with the mission of your institution and how will they support you?

Your application needs to briefly but clearly describe an innovative or novel problem, hypothesis, or challenge; a situation where we need to do better.
(Christian, 2018, P.36)

The abstract and specific aims are the two most important stages in the proposal. One or the other of these pages may be the only part of the proposal that some of the reviewers will read.
The specific aims should be written first to fit within one page and then trimmed as necessary to fit within the abstract box and augmented with brief statements of significance and experimental methods.
Reviewers have their own styles of review, but most probably start by quickly scanning the specific aims. Thus, this section also has a major impact on the primary reviewers and ultimately on the priority scores.
Preparation of a research proposal should start with the specific aims; the rest of the proposal merely amplifies what is presented there. After reading a well written specific aims, an experienced reviewer will understand the problem addressed, the hypothesis being tested and the feasibility and power of the experimental approach and will have a feeling about their importance.
(Ogden, Goldberg, 2002, P.55,56)

In Volume 2 of this series, we briefly touched on the concept of SMART objectives. In this Volume, we will cover it in more detail.

A goal can be defined as the result of a series of objectives that must be reached, and they themselves can be divided into a series of sub-objectives.

As for the criteria, they are the elements necessary for establishing a judgment, while the indicators are used to verify that they are met. Thus, a criterion for defining the concept of time for the execution of an objective can be controlled by a time indicator, such as "in one week". Each of the five elements that make up the SMART criterion is outlined below:

Specific: The objective must relate to a specific element. This criterion avoids formulations that are too broad, and therefore too vague.
By precisely defining an objective, the actions needed to achieve it become clear. Sub-objectives may be added (decreasing the defect rate, the number of failures, etc).

Measurable: It is essential to define the means to measure the level of achievement of the objective.

Assignable: One or more people should be clearly identified as responsible for the realization of the objective. These can be internal or external collaborators.

Realistic: This concept aims to differentiate the ideal situation, more difficult to achieve, from the concrete objective. It must be possible for the objective to be achieved with the current or new reasonably accessible means. In setting the objective, the legislation in place must also be considered for it to be realistic. This criterion will have an impact on the motivation and involvement of employees, so that it also strikes a balance between a challenging objective and an achievable objective. It may be useful to envisage another less ambitious objective in the case of failure.

Time-bound: it is important to define a time-line when setting an objective. Without time markers, the objective may actually lose its concreteness, and it would therefore not be possible to verify the achievement.
  ("50 Minutes", 2015, P.7,8)

Identify your key result areas, which come from your mission and what you offer to your customers. These will facilitate goal setting and developing corporate initiatives and strategic measures. With goals, established objectives can be determined. Making them smart objectives facilitates measurement.
(T. Howell, 2006, P.21)

Relation to policies, and work program

Policies

When preparing a project proposal, it is important to ensure that it will meet the policies and regulations that will govern it, if funded. The policies that should be considered include national policies, European Union policies, international policies, etc.
If, for instance, the project is being undertaken in Europe, it will need to comply with the policies of the European Union when seeking funding, so it is important to firstly identify the exact market that the project is operating in so as to know which policies

apply. The European Union has 283 different policy areas, one for each market; these range from "adult learning" to "zootechnics".
("ec.europa.eu")

Work program

The work program which the project proposal is being submitted through should be studied in great detail before any submission is made. By understanding the requirements and goals, you will be more equipped in aligning your project and proposal with the work program.
For instance, according to the European Commission, the EU will invest EUR 23.2 Billion, between 2014 and 2020, to co-fund TEN-T projects in EU Member States. Since 2014, the first CEF programming year, there have been four yearly waves of calls. In total, CEF has so far supported 641 projects with a total amount of €22.3 billion.
(ec.europa.eu)

Taking this programme into account, it is necessary to align the impact of the project with these aims of the European commission so that the project being submitted for funding is of relevance to the EU's goals.
(European Commission).

Concept and methodology

1. Background and Concept of the project

This section answers the questions as to **why** the project is needed and **what** the project is developing to meet the identified need.

Justify the scientific importance and interest of the research. Because you have probably thought a lot about the research you are proposing, it may be obvious to you why the research is important. But, do not expect it to be obvious to the reviewers of your proposal. You have to justify to them the importance of the research. Do not assume that others will see this importance without you stating it. An ineffective argument for the importance of research is to point out that X, Y, and Z have been done, but A, B and C have not yet been done and your goal is to do A, B and C. The fact that something has not been done does not, in itself, make that thing important. You need to show why your particular set of studies is worth doing.
(Sternberg, 2013, P.19)

This is the section where you have the opportunity to extensively describe in detail the *ideas underpinning the project's concept* and support them with *literature references.*
If it applies, you will also emphasize the *inter-disciplinarity of the project* by describing the *stakeholder's expertise.* A table listing the project-related *national or*

international research and innovation activities of the stakeholders will be determinant to demonstrate the highest level of support.

This section is also used to position the project in the Technology Readiness Level (TRL) scale, which describes the expected level of development of the technology during the project. The TRL scale spans from 1 to 9. Its levels correspond to the following development stages (European Commission):

- TRL 1 – basic principles observed

- TRL 2 – technology concept formulated

- TRL 3 – experimental proof of concept

- TRL 4 – technology validated in lab

- TRL 5 – technology validated in relevant environment (industrially relevant environment in the case of key enabling technologies)

- TRL 6 – technology demonstrated in relevant environment (industrially relevant environment in the case of key enabling technologies)

- TRL 7 – system prototype demonstration in operational environment

- TRL 8 – system complete and qualified

- TRL 9 – actual system proven in operational environment (competitive manufacturing in the case of key enabling technologies; or in space)

2. Methodological approach used by the project to achieve the objectives

This section answers the question as to **how** the project is going to be executed

The most important and lengthiest section of a project plan or proposal is the description of the methodology. The methodology must describe in detail how the project will be carried out. The methodology section should include the following:

1. A statement of the general methodological area. This aspect doesn't need to be elaborate, but it is essential to convey a clear understanding of the general domain (historical, descriptive or experimental) within which the project lies. This may be a simple statement such as "an experiment will be conducted to determine whether structured training in the use of internet search engines is associated with the quality of resources chosen for research papers in a freshmen level English composition book". The statement of the methodological area (experiment) is closely tied to the specific nature of the project and provides a preview of the remainder of the methodology.

2. If applicable, identification of the population to be studied. The population for a study consists of all those entities that are actually of interest to the project. The population is defined primarily by the nature of the problem, but requires specific definition within the plan or proposal.

3. Selection of a sample. If the above is applicable, the sample for a study consists of those members of the population that will actually be studied. The sample is defined primarily by the project designer in an effort to establish a manageable approach to studying a population. The essential goal of the sample selection process is to identify a sample that is adequately representative of the population.

4. Definition of terms, including both conceptual and operational definitions. Operational definitions need to be carefully and explicitly stated. Some terms can be assumed to be generally understood and not in need of definition, but the author should reach that conclusion carefully, not casually. If the meaning of a term is in any way modified for purposes of the project, the operational definition of that term must be stated.

5. Delineation of assumptions related to the project.
The fundamental principle of quality research or evaluation is that assumptions are known, articulated and justified. The author has an explicit duty to share key assumptions with the reader of the plan or proposal as a means of conveying a shared understanding of the nature of the proposed project.

6. Description of a data gathering plan.
The data gathering plan describes in detail the ways in which data will be acquired for the project. It includes a description of the origin and nature of the tools used to gather data (observation, instruments, interview, questionnaire, etc) anticipation of any barriers to effective data gathering and discussion of the ways in which data will be measured.

7. Description of a data analysis plan.
The data analysis plan describes in detail the impact of missing or faulty data, the approach that will be taken to exploring the relationship of the primary data to context data and to exploring relationships within the primary data and the ways in which significance will be assessed.

(Wallace,Van Fleet, 2012, P.120,121)

Ambition

In this section, you will explain how the project enables moving beyond the state of the art. Usually this is achieved by explaining all possible innovations developed and implemented during the project.

There are numerous definitions of innovation, but generally innovation refers to a new product, idea or process which introduces a significant positive change that adds value to the organization.
(Schneider, Fuller, 2018, chpt.6)

When preparing a project proposal, the innovative approach must be clearly outlined to show how that particular project is introducing a new way of identifying/solving the issue being addressed.
It is not sufficient to justify the projects goal by stating that this issue/area has been addressed before. There must be a novel/innovative aspect that allows that particular project to progress the issue being tackled in a way that no other, previous project or research effort has been able to do to date.

Tips

- Clearly describe the novel problem; the situation where we need to do better.

- Identify the exact market the project is operating in, to know which policies apply.

- Do not assume that others will see this importance without you stating it

Expected impacts

Research impact is defined by UK economic and social research council (ESRC) as "the demonstrable contribution that excellent research makes to society and the economy". This can involve "academic impact", "economical and societal impact" or both. ESRC explains that academic impact is "the demonstrable contribution that excellent research makes in shifting understanding and advancing scientific, method, theory and application across and within disciplines."; economical and societal impact is "the demonstrable contribution that excellent research makes to society and the economy and its benefits to individuals, organizations and/or nations."

The impact of your project can be assessed at different levels, for example science, industry, public, patients, policy makers, society, economy, both local and national, EU and global.
Understanding your impact will allow you to demonstrate it to others. Be aware that your research impact is wider than your count of citations. Or it should be. While you aim for publications, and for those publications to be read and then cited, the impact of your research will be felt beyond those publications as it adds to knowledge in your field and/or influences changes in practice.
(Christian, 2018, P.373)

When producing the project proposal for submission, it is important to show that the project's potential impact matches the expectations of the call of the funding agency.

Measures to maximize the impact

1. Dissemination of results

Evaluation and dissemination are the factors that give an indication whether or not the project has been successful. Although evaluation and dissemination are carried out at the end of the project, they need to be integrated into the project plan from the start.
("Entrepreneur Press", 2012, P.78)
Dissemination refers to the distribution of information. Funding sources frequently pay for disseminating the results of your grant project through one or more of the following mechanism:

A report mailed to others in the field,
A quarterly journal,
A newsletter,
A seminar or conference on the topic,

Participation in national conferences,
A film, CD or DVD presentation on the project,
A website.

Dissemination helps to spread word of the project. It helps to increase public awareness of the program or project, pave the way for soliciting additional support, let others in the field know of the research findings (in the case of a research grant) and contribute to available information.
Assuming that the project is successfully funded and reaches completion, dissemination may be the next step for the project. If the results are going to be made available by dissemination, there are multiple factors to take into account when writing the dissemination, such as:

- Clearly identify the intended results of the dissemination effort: the intended results of a project may include a journal publication, video documentary or a website among other resources.
- Specify precisely who will be responsible for dissemination and the people's qualifications.
- Discuss internal and external project dissemination.

("Entrepreneur Press", 2012, P.81)

Funding agencies, in keeping with their usual emphasis on supporting projects that are broadly applicable, tend to expect grant recipients to actively seek to disseminate results beyond the required final report to the funding agency. Research projects don't always lead to publications as methods of dissemination, but publication is a frequent goal of a research project. Leaders of internal evaluation projects must plan for dissemination of results to administrators, decision makers and other essential stakeholders.
(Wallace, Van Fleet, 2012, P.123)

2. Exploitation

Commercialization is the process of managing the transfer of research knowledge to the place where it becomes an application in the broad marketplace. The knowledge might be a research outcome or skill; it might result in the development of a product, a technology, service or business, a community development program or consulting activities. When applicable, the project description will contain a value chain graph, representing how the progressive achievement of the different objectives of the project increases the value of the product.

The idea of the value chain is based on the process view of organisations, the idea of seeing a manufacturing (or service) organisation as a system, made up of subsystems each with inputs, transformation processes and outputs. Inputs, transformation processes, and outputs involve the acquisition and consumption of resources -

money, labour, materials, equipment, buildings, land, administration and management.

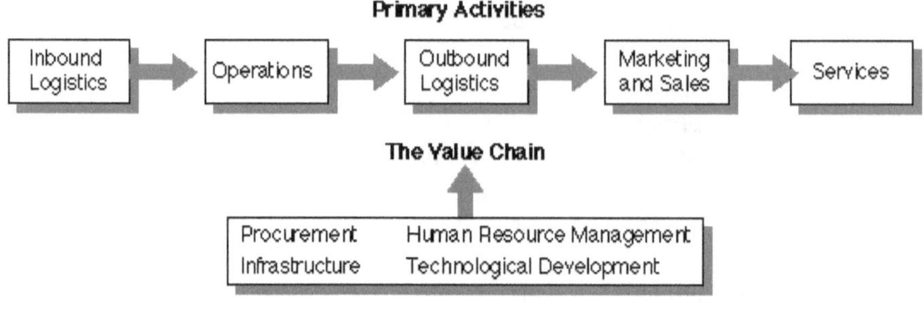

(University of Cambridge)

Some ways to protect your IP for commercialization might include the following types of registrations: design registration, registered trademark, copyright law. (Christian, 2018, P.384, 386, 387)

The description of the commercialization plan is best achieved through a business plan. At least six key elements should be described in such plan:

A. *Proposition general description*
    A. *Type of exploitable result*
    B. *Components and value*
B. *Clients*
    A. *Customer segments*
    B. *Customer relationships (Get, Keep, Grow)*
    C. *Addressing the ethical issues of the customer*
    D. *Channels of communication*
C. *Competition*
D. *Financial aspects*
    A. *Revenue streams*
    B. *Costs*
E. *Concerns before setting up a startup*
    A. *Key activities*
    B. *Key resources*
    C. *Key partners*
F. *Validation of the business model (regular review)*
    A. *Problem-solution fit*
    B. *Product-market fit*
    C. *Business model fit*

Aside from putting together a business plan, it is also beneficial to form a business model.

A business model is a simple representation of the complex reality of a business. The primary purpose of a business model is to communicate something about the business to other people: employees, customers, partners or suppliers.

A good business model supports different views of the same underlying knowledge. Each subject matter expert can see what they need to see, for their own purposes. Each can ignore the detail needed for other subject matter experts.

For example, a strategist can look at an organizations goals, strategies and tactics, ignoring the business processes and interactions. A sales specialist can examine the business processes supporting sales, ignoring the processes supporting operations and maintenance.

Business is a communication intense activity. Some of that communication is complex. For example, business policies change. Business models are better for conveying complex business information.

Business models are effective for communication because most models are visual. Diagrams make a model easier to understand and faster to communicate.

(Bridgeland, Zahavi, 2008, P.1,10,11)

Below is an example of a business model canvas.

Shopkeeper Example	Advanced Business Model Canvas			Key Attributes: Have *Want*
Key Partners	**Key Activities**	**Value Proposition**	**Customer Relationships**	**Customer Segments**
Distribution partners Transport providers Component suppliers Industry Associations Tech. suppliers Tech. maintenance provider *Remote VOIP telephony* *Digital analytic providers* *IT remote support providers*	Sales Packaging Logistics Procurement Accounting	Quick service Accessible location Convenient Digital payments	In premises one to one *Self service* *In store demonstations* *In store video* *In store QR codes for more info.* *Loyalty card with analytics* *Social media interactions*	Niche market Retail Buyer not always end user Customer needs are satisfied
	Key Resources People Store Power Communications Other Utilities Sales data *Customer data* *Decision support systems* *Industry analytics* *Knowledge bases*		**Channels** Leaflet distribution Sponsored groups Signage *Website* *Internet map search* *Web page product awareness* *Web page how you help* *Smart digital signage* *Customer feedback system*	
Cost Structure		**Revenue Streams**		
Labour Taxes Building expenses Communications Printing Insurance Waste removal Energy Sources	Franchises fees Technologies *Web hosting* *Content creation* *Pay per click ads* *Photos, video, audio, illustrations*	Product sales Accesories sales *Renewable energy creation*		

(www.matthewb.id.au)

During the course of the project, excellent communication between the project stakeholders will be key to achieve the goals.

The communication strategy will define with whom to communicate (or not) and about what, when and how. The plan will also define what are the information needed by each player to be operational.

Tips

- *Read the work program and address each expected impact*

- *Be realistic in formulating the impact*

- *Impact means what remains of the project after the funding period*

Defining Work packages

As described in volume 2, work packages will be defined as a group of tasks that enable the achievement of specific objectives.

Each work package will therefore be defined by:
- Two or three specific objectives;
- A description of the work and the role of the work package participants;
- The description of the tasks, including their aim, duration, and dependencies to other tasks in other work packages;
- The deliverables, corresponding to the outcomes of the work package (these can be reports, events, publications, prototypes, etc).

Generally, work packages are used to separate activities by types within the project, such as management, technical, outreach, exploitation, legal activities, among others.

PERT Chart

PERT (Program Evaluation and review Technique) charts can be used to structure the work of the project. The chart consists of a diagram that presents the sequence of activities which need to be completed to ensure the implementation of a project or program. As exemplified in the figure, PERT charts are generally structured as follows:

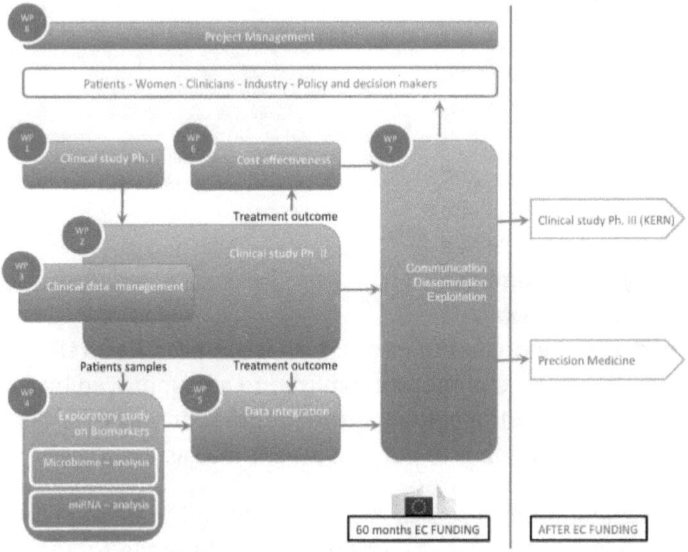

- The time frame moves from left to right,

- Boxes represent work packages or tasks,

- Arrows are activities or deliverables that connect work packages and must be performed in order to enable the next work or task,

(Timmreck, 2003, P.153,155)

Gantt chart

The Gantt chart is based on action - doing, getting projects or activities done. A heavy horizontal line is drawn in each row to show the amount of time each person, project or activity needs to complete the task. Every heavy line is connected with an arrow to another heavy line to indicate the interdependency between the tasks.

The Gantt chart will be used during the project execution for the project manager to follow the progress of the tasks and identify delays and their downstream consequences.

When many different activities are to be conducted for one project, several Gantt charts can be developed - one chart for each activity could be developed. These could be compared with each other to check progress and coordinate the various activities.

(Thomas C. Timmreck, 2003, P.152,153)

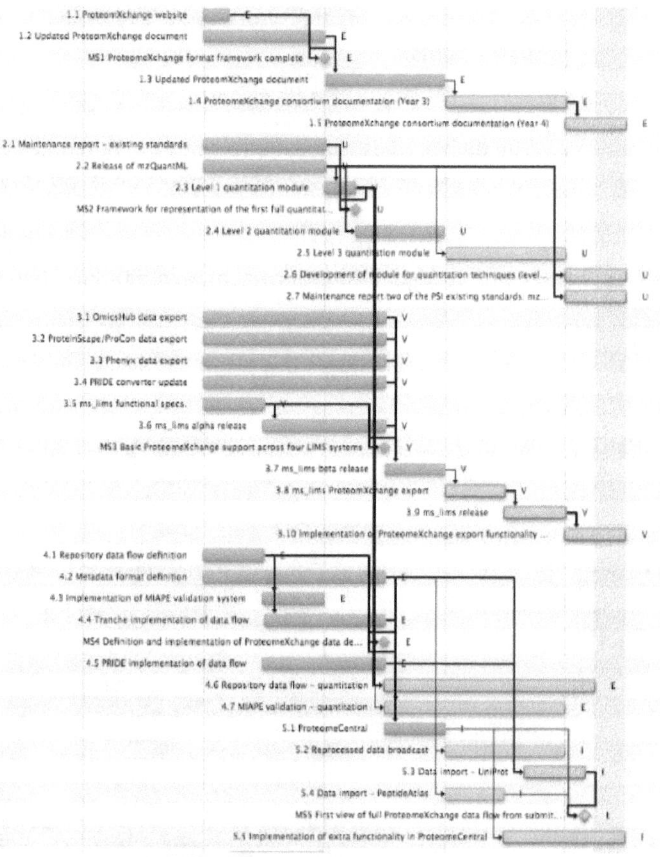

Project's success depends on how its organization is communicated and how its tasks are understood and carried out by its members. The manner in which a project is to be organized in order to accomplish its intellectual goals will likely first be described in the proposal submitted to and approved by a funder.

Whether you are the author of the proposal or you have been recruited to manage the project, your job is to put those plans into action.

Although different research methods imply different approaches to management, it is a managers strategic responsibility and challenge to adapt an approach, or more likely to find a mix and balance between approaches that will best keep the particular members of a research team in sync with one another, on track with the projects specific schedule and budget, and in support of what are likely to be a projects distinctive, perhaps unique, research goals and questions.

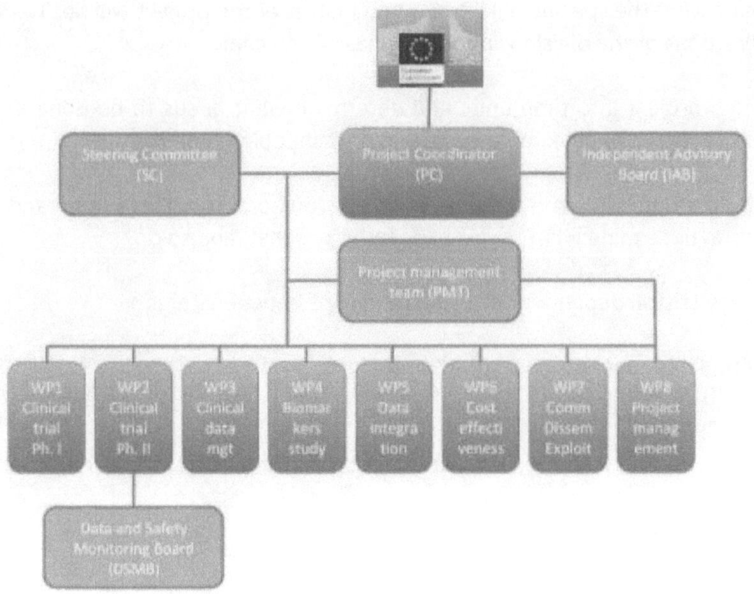

The above diagram is an example of project management structure.

As can be seen, the project management team is responsible for the monitoring of the individual WP'S (work packages). The work package is the critical level for managing a work breakdown structure. Each WP has one or two leaders, usually experts in the field tackled within the WP, who report to the project management team. A Steering Committee can be formed by all or by some of the WP leaders, in order to advise the project coordinator on progress and possible decisions to be made. The project coordinator can also seek advise by an Independent Advisory Board, composed by persons external to the project, such as industry leaders, experts in the field of research, policy makers, etc.

The management structure should also connect the funding body to the project coordinator or to the entity that will report on the execution of the project.

(Kerzner, 2009)

Milestones

A combined analysis of the PERT diagram and of the Gantt chart can help determining what the specific milestones and phases of the project will be. To define these milestones or the divisions between phases, you can:

  - break the project into time units and determine what needs to be done in each unit of time in order to be completed within the scheduled timeframe.

- start with the outcomes and the date by which you hope to achieve them and work backwards to determine what has to happen before that, and so on.

- identify any critical dependencies or relationships between items.

- use a critical path analysis chart or a Gantt chart.

("Bookboon"P.30)

Risk

Risks are events, circumstances, situations or conditions that might occur with a specific probability and have a potential negative impact on meeting predefined project objectives. Risks are the so-called "known unknowns" and can be separated from uncertainty if they are measurable, calculable and predictable for the course of the project.
(Bodea, Nicoleta, 2016, P.3)

When preparing the research proposal, it is important to do a risk analysis. The steps to performing a risk analysis for the research project include;

> - identify all events potentially generating risk for the project including their origin, nature and consequences.
> - Evaluate each risk by assigning a probability of occurrence and a factor of impact.
> - Stratify risks in order to concentrate the attention on major risks.
> - Develop a plan of action to limit the impact of major risks.
> -   Follow-up and update risks regularly during the project.

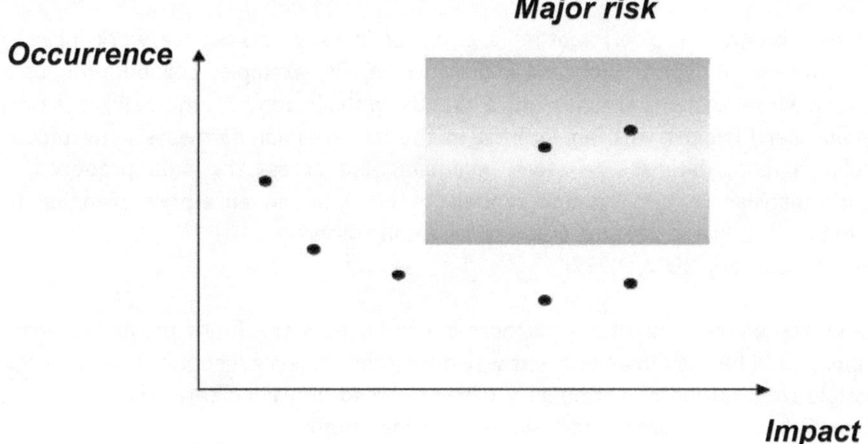

There are many factors to take into account when performing a risk analysis for a research project, such as;

-Human factor
-Technical factors
-Financial factors
-Environmental factors
-Factors linked to customers
-Historical factors
-Organizational factors
-Cultural factors

Consortium

A common form of partnership seen in the academic research environment is the research consortium. Consortia may also be structured as separate corporate entities and sometimes as loose alliances with only contractual agreements to serve as the structural framework.

The most important issue to be addressed in establishing a consortium is the allocation of control over the consortium. Organizations establishing consortia need to establish just how much freedom the consortium will have to take risks and redirect research as circumstances change in order to achieve the project's objectives.

(Kulakowski, Chronister, 2008, P.361, 362)

The best program projects feature a group of investigators whose work logically interconnects. A typical program might involve, for example, collaboration of a clinician whose patients are diabetic, a scientist with a colony of atheroelerotics rats, a molecular biologist with an interest in the intra-cellular pathways activated by insulin, a large database to store, maintain, and access the data produced, a bioinformatician integrating and modelling the data, an enterprise planning to exploit the product, a patients association, among others.
(Ogden, Goldberg, 2002, P.237)

The choice of the consortium partners is key because the funds provided by the grants should be used in an efficient way during the project execution. Therefore the possible contribution of a candidate partner should be valued carefully against the scope of the project, before adding it to the consortium.

The funding body usually describes in its call the scope of the project, if particular disciplines are expected, or recommendations in the choice of participating countries.

Tips

- Remember, the success depends on how the organization is communicated and how its tasks are understood.

- The risk analysis must be clearly outlined in the proposal.

- Determine what the specific milestones and phases of the project will be

Resources distribution

A carefully prepared and documented budget is perhaps the most obvious element of any proposal or plan. The budget explains to those individuals responsible for approving the project exactly what monetary and non-monetary investments will be necessary for completion of the project and how those investments will be funded. Every project, whether internally supported or dependent on external funding, is associated with costs that require appropriate and complete accounting. Applications for external funding must describe the need for funding in explicit and thorough terms and detail either all costs of the project or only those costs for which funding is actually sought. For either internally or externally funded projects, it is important for internal purposes to carefully analyse, document and budget all costs associated with the project. The following are prominent budget lines for research or evaluation proposals:

1. Personnel project personnel fall into three categories:

A. Permanent employees of the host institution who will devote some portion of their time to the project. It is typically the case that permanent employees devote only a portion of their time to the project and spend the remainder of their time in fulfilment of their regular duties. The grant budget may request funding to recover the employees wages for time devoted to the project.
The norm for a salaried employee is to calculate the percentage of the employees time that will be spent on the project and budget that percentage of the salary for the project period.
In addition to wages, it may be necessary to include a budget line for benefits for either permanent or temporary employees. Benefits are typically budgeted as a percentage of wages, but may be determined more precisely by calculating the actual costs of insurance, retirement, workers compensation and related benefits.

B. Temporary employees who will be hired for the project. Depending on the nature of the project, it may be necessary to hire technical staff, clerical staff, computer staff or student employees whose sole employment responsibility is to the project.

C. Consultants who will be paid on a contractual basis to perform specific tasks. Consultants are typically paid a flat fee negotiated individually with each consultant.

2. Facilities. A realistic assessment of the need to budget for facilities is an essential component of the process of developing a research or evaluation proposal. The most prominent facilities issues have to do with space, furnishings and equipment.

A. Space. Space is frequently a cost absorbed by the host institution, although long-term complex projects may require rental of office, laboratory or other space. Budgeting for dedicated use of existing space, can also be difficult, but can be accomplished by determining typical renting rates in the community.

B. Furnishings. Access to furnishings is often overlooked in the preparations of research or evaluation proposals.

C. Equipment. The most frequent need for equipment involves computer hardware and software.

3. Materials, supplies and services. Unit costs for materials, supplies and services are usually fairly easy to determine, but doing so can be time consuming and should not be left until the last moment.

4. Travel. Some research or evaluation projects require travel to gather data or for other purposes related to successful completion of the project. One of the difficulties of budgeting for travel is that some components, such as airfare, may change between the time when the proposal is developed and the time when the travel actually takes place.

5. Support services. Some research or evaluation projects require contractual support services, such as special custodial services, transcription of audio recordings, use of centralized computing facilities, etc.

6. Overhead/indirect costs. Indirect costs are the overhead items that are necessary to support the institution in which the project is to take place.

7. Justification of budget items. Many funding agencies require, in addition to a budget presented in a line-by-line format, a narrative discussion of the need for and nature of the items included in the budget.

(Wallace, Van Fleet, 2012, P.124-128)

Funding rate

When putting together the budget section of the project proposal, one important aspect is the funding rate and reimbursement percentages.
The burn rate of the project should be precisely calculated along side the percentage of costs which will be covered by grants, investment, etc, in order to understand where all required finances will come from and what, if any, additional financing will be required.

- Describe the need for funding in explicit and thorough terms

- Analyse, document and budget all costs associated with the project

- Make sure to Justify all budgeted items

1. SMART Criteria: Become more successful by setting better goals by 50 Minutes (2015) P.7,8

2. Actionable Performance Measurement: A Key to Success by Marvin T. Howell (2006) P.21

3. European Commission - "https://eur-lex.europa.eu/summary/chapter/regional_policy.html?root_default=SUM_1_CODED=26"

4. Start Your Own Grant Writing Business: Your Step-By-Step Guide to Success by Entrepreneur Press (2012) P.78,81

5. Knowledge into Action: Research and Evaluation in Library and Information Science: Research and Evaluation in Library and Information Science by Danny P. Wallace, Connie J Van Fleet (2012) P.123

6. Writing Successful Grant Proposals from the Top Down and Bottom Up by Robert J. Sternberg (2013) P.19

7. Research Proposals: A Guide to Success by Thomas E. Ogden, Israel A. Goldberg (2002) P.55, 56

8. Knowledge into Action: Research and Evaluation in Library and Information Science: Research and Evaluation in Library and Information Science by Danny P. Wallace, Connie J Van Fleet (2012) P.124-128

9. Planning, Program Development, and Evaluation: A Handbook for Health Promotion, Aging, and Health Services by Thomas C. Timmreck (2003) P.153,155

10. Total Quality Management: The Key to Business Improvement by C. Hakes (1991) P.90

11. Building the European Research Area: Socio-economic Research in Practice by Michael Kuhn, Svend Remøe (2005) P.273

12. Project Management: A Systems Approach to Planning, Scheduling, and Controlling by Harold Kerzner (2009) P.437

13. Managing Projects by "Bookboon", P.30

14. Keys to Running Successful Research Projects: All the Things They Never Teach You by Katherine Christian (2018) P.36, 384, 386, 387

15. Writing Research Proposals in the Health Sciences: A Step-by-step Guide by Zevia Schneider, Jeffrey Fuller (2018) Chapter 6

16. https://ec.europa.eu - European Commission

17. (UniversityofCambridge, "https://www.ifm.eng.cam.ac.uk/research/dstools/value-chain-/")

18. Business Modeling: A Practical Guide to Realizing Business Value by David M. Bridgeland, Ron Zahavi (2008) P.1,10,11

19. www.matthewb.id.au - https://www.matthewb.id.au/b/business-model-examples.html

20. The SAGE Handbook of Research Management by Robert Dingwall, Mary Byrne McDonnell (2015) P.192,193

Evaluation criteria

Proposal guidelines must be followed. These are different for each different funding program. Any deviations from page restrictions, bio-sketch format, section order, IRB or IACUC requirements, etc, may result in a worse priority score or even cause the proposal to be returned without review.
Reviewers expect a specific order of presentation and length.
Although the assigned reviewers will read the proposal in detail, most of the remaining reviewers will merely flip through it looking red flags.
The most important of these are found in the budget and in the bio-sketch. The budget must be reasonable at a glance. The reasonable threshold varies greatly among institutes and different types of research.
The bio-sketch should indicate, also at a glance, solid training, steady productivity, and recent publications pertinent to the proposed research.
The reviewer must be given the impression that the investigator is absolutely capable of carrying out the proposed research and that there is a high likelihood of success and significant publications.
The specific aim section of the research plan is the most critical page of the entire proposal. Failure of the reviewer to understand the specific aims presages disaster for the review and is always the fault of the investigator. It is an insurmountable red flag.
Diagrams are a must. A well designed diagram in the specific aims or the background and significance sections may reveal at a glance general theory, what is already known and hypothesis and their tests.

(Ogden, Goldberg, 2002, P.23,24)

It's not a scientific publication, it's a selling document

Unfortunately, good science does not always lead to a successful proposal, although badly written proposals are often funded if their science and the principal investigators background are sufficiently strong. On the other hand, proposals based on faulty science are hardly ever successful. Between these extremes lies a group of proposals whose science is sound and with principal investigators that are well trained and productive. Some will be funded, others will not.
A well written proposal is written to communicate with all the reviewers, not just those with expertise in the field.
(Ogden, Goldberg, 2002, P.21)

As mentioned in volume 3 of this series, the abstract and specific aims are two of the most important pages in the proposal. One or the other of these pages may be the only part of the proposal that some of the reviewers will read.
The abstract should contain the essence of the specific aims, a few short sentences concerning the health relatedness of the research and its scientific significance in terms of its long term goals.
(Ogden, Goldberg, 2002, P.55,65)

Abstracts are supposed to be shorter than their originals. This shows that the length of the abstract is a question of prime interest.
Several sources have advice on the matter of length. Day (1988), for example, votes for a maximum length of 250 words. The same measure is set in the 1977 ANSI standard. Its ads, however, that the abstracts length should be appropriate to the potential usefulness of the document abstracted (ANSI, 1997).
Waters (1982) is of the opinion that in abstracts shorter than 200 words important information may be missing, while abstracts over this limit could contain much redundant information. He adds, notwithstanding, that the originals length is decisive.

Generally speaking, the purpose of abstracting is to give an idea about the content of a text to someone who does not know that text. This is done without retelling every detail of the original (Werlich, 1988).

The main functions of abstracts are expressed by Fidel (1993) who states that abstracts increase the efficiency of information gathering as they:

-Give orientation to users,
-Provide an overview for those who need to keep up to date,
-Serve as a source of information

Cleveland and Cleveland (1983) state that good abstracts avoid both the bias and personal viewpoints that may be introduced by critical comments.

(Koltay, 2010, P.33,34,36,37,54)

Its all about structure

The gold standard of proposal writing is exemplified by any article in the *Scientific American*. These are written for readers who are scientists but are unfamiliar with the area of the particular article. The prose is kept simple, specialized words and abbreviations are avoided and every page has at least one diagram or figure.
Every page of a proposal should have the same general appearance.

It is poor grantsmanship to use different fonts or to insert obviously photocopied pages from other proposals. It makes a bad impression, for instance, to include a bio-sketch photocopied from a previous submission. This is a common failing of collaborative projects in which the co-pi hands the PI a bio-sketch prepared several years previously, obviously for a different purpose. This indicates to the reviewer that the collaborator does not take the project seriously enough to update the bio-sketch. It speaks poorly for the success of the collaboration.

Good writing is brief

Reduce the text sufficiently that the page limitations can be met with at least a size 12 font. Double space between paragraphs and use 1.2 line spaces between lines. Use diagrams to reduce narratives and paragraph titles to facilitate skimming.
A strong proposal will give the appearance of being well organized and readable at first glance.

(Ogden, Goldberg, 2002, P.21,22,23)

II. Final Tips

- A grant proposal is NOT a scientific publication, it is a selling document

- Write in simple words

- The abstract page is critical (PITCH)

- Structure your text in short paragraphs

- Give a logical structure

- Use graphics and tables when appropriate

- Imagine that you are the evaluator

- Use templates as a guideline

- Check the evaluation forms

1. Research Proposals: A Guide to Success by Thomas E. Ogden, Israel A. Goldberg (2002) P.21,22,23,24,55,65

2. Abstracts and Abstracting: A Genre and Set of Skills for the Twenty-First Century by Tibor Koltay (2010) P.33,34,36,37,54